JELLYFISH

LIVING  THINGS

JELLYFISH

Rebecca Stefoff

BENCHMARK BOOKS

MARSHALL CAVENDISH
NEW YORK

Benchmark Books
Marshall Cavendish Corporation
99 White Plains Road
Tarrytown, New York 10591-9001

Illustrations by Jean Cassels

Library of Congress Cataloging-in-Publication Data
Stefoff, Rebecca, date
Jellyfish / Rebecca Stefoff.
p.   cm. — (Living things)
Includes bibliographical references and index.
Summary: Examines the physical characteristics and behavior of jellyfish
and describes six different kinds.
ISBN 0-7614-0411-2 (lib. bdg.)
1. Jellyfishes—Juvenile literature. [1. Jellyfishes.]
I. Title. II. Series: Stefoff, Rebecca Living things.
QL377.S4S74 1997   593.5'3—dc21   96-39107   CIP   AC

Photo research by Ellen Barrett Dudley

Cover photo: *The National Audubon Society Collection/Photo Researchers, Inc.*,
Gregory Ochocki

The National Audubon Society Collection/Photo Researchers, Inc.: Tom McHugh,
2; Verna R. Johnston, 7; Neil G. McDaniel, 9; Andrew J. Martinez, 10; Nancy Sefton,
12 (top); William Curtsinger, 12 (bottom); Charles V. Angelo, 15; Noble Proctor, 24;
Klaus Hilgert/Okapia, 25; Eric Haucke, 32. *Peter Arnold, Inc.*: Fred Bavendam, 8
(left), 14 (top), 23 (top); Norbert Wu, 8 (right), 13, 21; Kelvin Aitken, 14 (bottom);
Bob Evans, 17; Steve Kaufman, 26–27. *Animals Animals:* Herb Segars, 6, 19;
Kathie Atkinson, 11; Oxford Scientific Films, 16, 22, 23 (bottom); Fred Bavendam,
18; Lewis Trusty, 20.

Printed in the United States of America

3   5   6   4   2

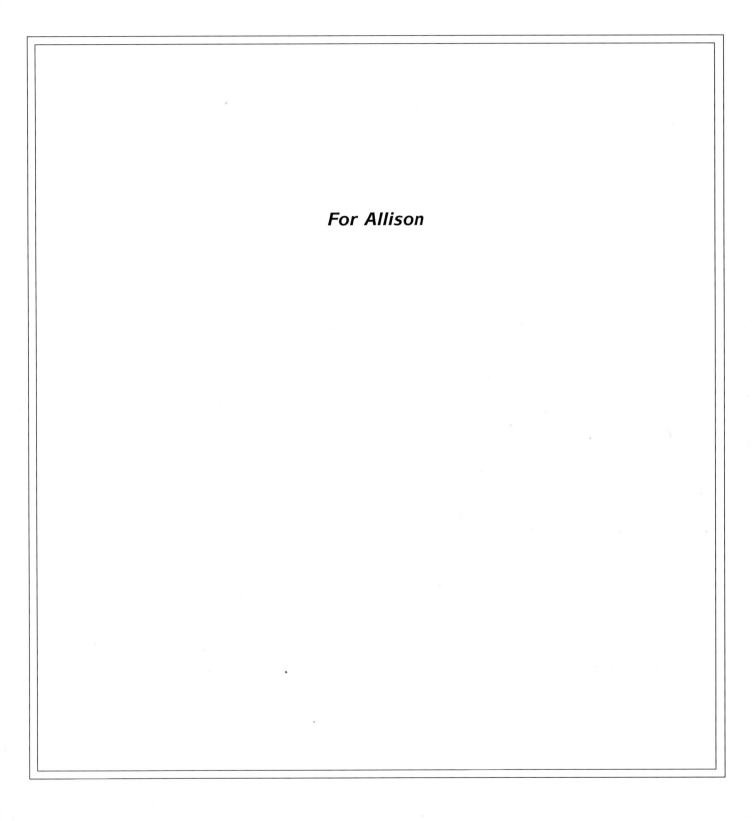

For Allison

cannonball jellyfish on beach

jellyfish and brown kelp

Have you ever walked along a beach and seen a jellyfish lying on the sand? The wind and the waves have carried the jellyfish to shore, but its home is out in the ocean.

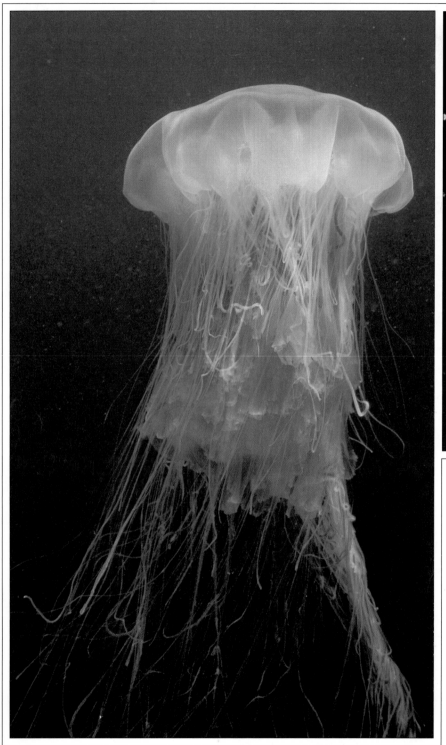

The jellyfish is not a fish. It got its name because its body is shiny and smooth and squishy, like jelly, and it lives in the sea, like a fish.

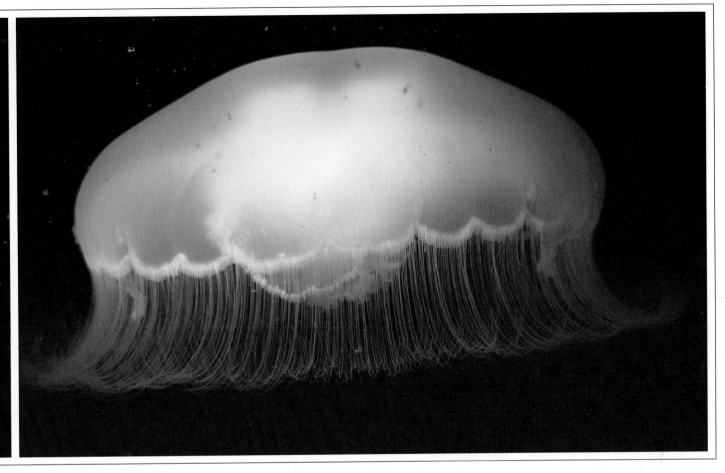

Most jellyfish are shaped like umbrellas, or upside-down bowls. They have arms called tentacles. The tentacles hang down in a fringe. Some tentacles are short and stubby, and others are long and thin, like silvery hairs. A few jellyfish have tentacles one hundred feet long.

Jellyfish swim by squeezing water through their bodies, but they can't swim very fast. Most of the time they just drift along, letting the sea carry them wherever it will.

Some jellyfish live at the top of the ocean. Their bodies stick up like sails, and the wind pushes them along as though they were little ships.

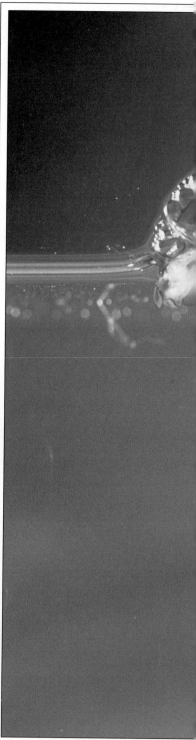

upside-down jellyfish

Portuguese men-of-war

Jellyfish names sometimes tell what the jellyfish look like. Have you ever seen a small, round-topped sewing thimble?

helmet jellyfish

If you have, you can guess which of these jellyfish is called a thimble.

The thimble jellyfish is one of the smallest jellyfish in the world. Its body makes its own light, so the thimble jellyfish glows at night or in deep, dark water.

The helmet jellyfish is shaped like an old-fashioned knight's helmet—but no knight ever wore those long pink tentacles! The glass jellyfish is as clear as a window. You can see right through it.

glass jellyfish

The jellyfish has many relatives in the ocean. Two of its relatives are the anemone and the coral animal.

white-spotted tealia anemone

Anemones look like plants, but they are animals fastened to a rock on the seafloor. Coral animals are tiny. It took thousands of them to build this big piece of pillar coral.

pillar coral

Portuguese man-of-war eating fish

Jellyfish eat through tubes that hang down among their tentacles. This jellyfish has caught a fish. The tentacles sting the fish and kill it so that the jellyfish can slowly eat it.

While the jellyfish is finishing its meal, a fish might come along and eat *it*. Jellyfish are the favorite food of some fish. And when a dead jellyfish falls to the bottom of the ocean, the anemones and starfish there will eat it up.

anemone and starfish eating dead jellyfish

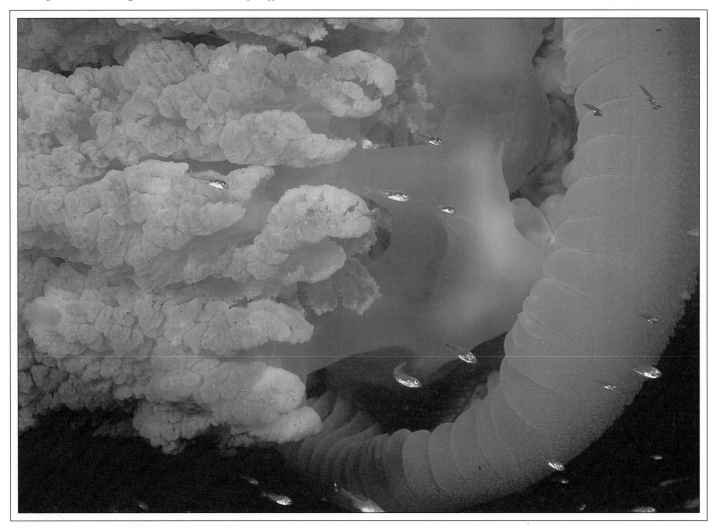

The jellyfish's stinging tentacles hurt most fish. Some jellyfish's tentacles would hurt you, too, if you brushed against them. But these little fish are not afraid. They have special skin. When the jellyfish's tentacles touch them, they don't get stung.

The fish are swimming right into the dangling tentacles. They know that no other hungry fish will chase them there. For these fish, the jellyfish is a safe shelter.

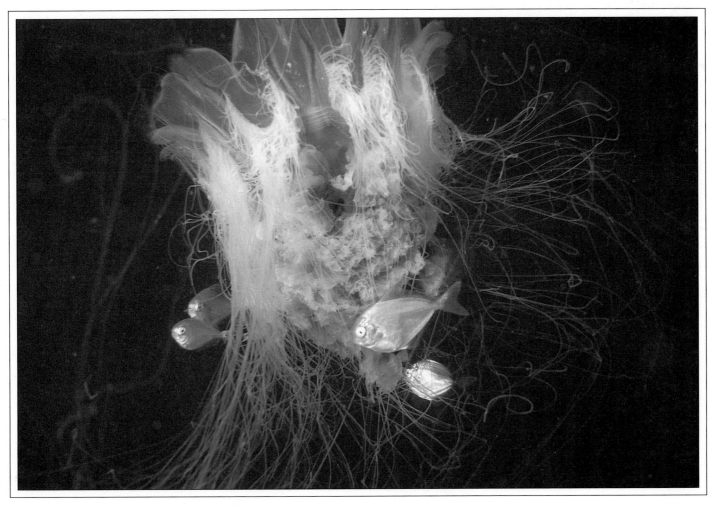

young butterfish and lion's mane jellyfish

medusa fish inside jellyfish

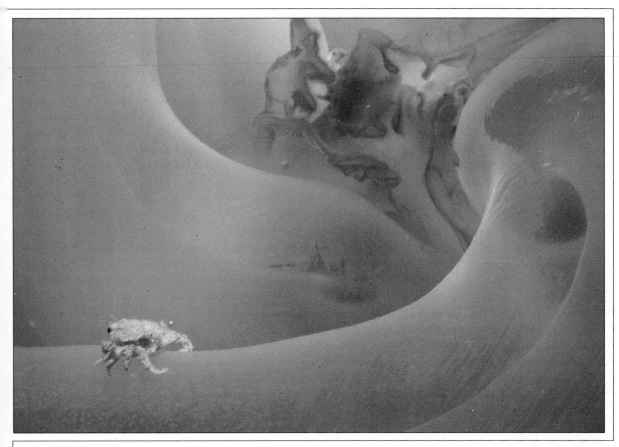

crab living inside jellyfish

Some fish and crabs live their whole lives *inside* a jellyfish. They eat scraps from the jellyfish's dinners and tiny plants that grow on the jellyfish's skin.

polyps attached to seafloor

Jellyfish start life as eggs floating in the sea. When an egg lands on something, it grows a stalk and stays fastened to that spot. Now it is called a polyp.

Some jellyfish, like the one fastened to the brown seaweed, stay polyps forever. But most polyps split up into tiny, free-floating animals that look like snowflakes. These little snowflakes of the sea grow up into the animals we call jellyfish.

young moon jellyfish swims free

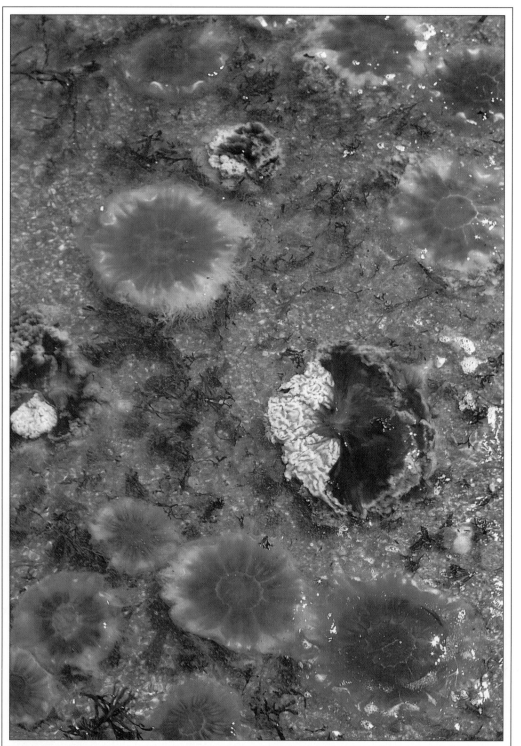

jellyfish washing into a tide pool

If you are lucky, you will see jellyfish in a tide pool someday. But if you are *very* lucky, you might see hundreds of them together out in the ocean.

Big clusters of jellyfish are called swarms. Sometimes a swarm drifts for miles and miles through the sea, like a living cloud. At night it glows like swimming stars.

moon jellies

All around the world, wind and waves wash jellyfish onto beaches. These jellyfish will die. They cannot live on land. But they remind us that our world's oceans are filled with life.

jellyfish on tidal flats, Alaska

A QUICK LOOK AT THE JELLYFISH

Jellyfish belong to a large group of animals called cnidarians (nih DARE ee uns). Nearly all jellyfish live in the oceans, but a few live in fresh water.

Scientists call jellyfish medusas. The medusa is the adult stage in the animal's life cycle. The cycle begins with a tiny fertilized egg floating in the sea. The larva attaches itself to something, grows a stalk, and becomes a polyp. A few kinds of jellyfish stay in the polyp stage, but most polyps break up into tiny free-swimming medusas. These grow into the animals we call jellyfish.

Here are six kinds of jellyfish along with their scientific names in Latin and a few key facts:

MANY-RIBBED HYDROMEDUSA

Aequorea aequorea

(ee KWOR yah ee KWOR yah)

Clear bowl-shaped jellyfish about seven inches across (17.5 cm). Floats in open water, sometimes near shore. Lives in all parts of the world. Often washes up on beaches. Glows at night.

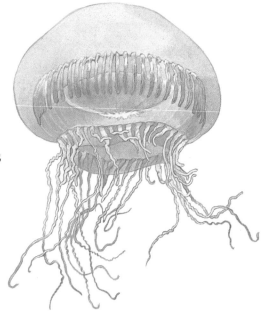

BY-THE-WIND SAILOR

Velella velella

(veh LEH lah veh LEH lah)

Measures about four inches across (10 cm), with triangular crest or "sail" two inches high (5 cm). Transparent blue. Floats on surface, pushed along by wind against "sail." Lives in warm parts of Pacific and Atlantic Oceans.

PURPLE JELLYFISH

Pelagia noctiluca

(peh LAG ee ah nock til LOO cah)

Generally about four inches wide (10 cm), with eight long, fine tentacles and a long, thick feeding tube. Most often purple or pink, but sometimes yellow. Glows at night.

MOON JELLYFISH

Aurelia aurita

(ow RAY lee ah ow REE tah)

Shaped like a saucer. Measures about sixteen inches across (40 cm). Tentacles are short. Sting causes slight rash.

BOX JELLYFISH

Chironex fleckeri

(kee ROH nex FLEH ker eye)

Also called sea wasp. Lives close to shore in waters of Australia and Southeast Asia. Body is about the size of a basketball. Tentacles may be fifteen feet long (4.5 meters). Sting is highly dangerous, often fatal. Experts say it is the most poisonous animal on earth.

LION'S MANE JELLYFISH

Cyanea capillata

(sigh ah NAY ah cap ih LAH tah)

One of the largest jellyfish in the world. Largest measure six or eight feet across (1.8–2.4 meters). Most are only one or two feet across (30–61 cm). Pink or yellowish when young, dark reddish brown when mature. Has about 150 long tentacles. Sting causes severe burning and blisters. Found in many parts of the world, including both coasts of the United States.

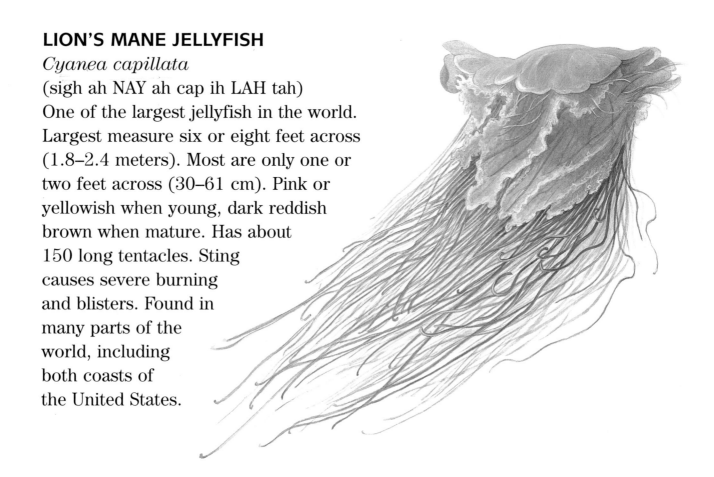

Taking Care of the Jellyfish

Like all animals and plants that live in the ocean, jellyfish are harmed by pollution of our ocean waters. Garbage, chemicals, and oil hurt all life in the sea. The best way to make sure that jellyfish will always be part of our world is to keep the oceans clean.

Find Out More

Gowell, Elizabeth. *Sea Jellies: Rainbows in the Sea.* New York: Franklin Watts, 1993.

Macquitty, Miranda. *Discovering Jellyfish.* New York: Bookwright, 1989.

Oxford Scientific Films. *Jellyfish and Other Sea Creatures.* New York: Putnam, 1982.

Shepherd, Elizabeth. *Jellyfish.* New York: Lothrop, Lee & Shepard, 1969.

Waters, John. *A Jellyfish Is Not a Fish.* New York: Crowell, 1979.

Index

Rebecca Stefoff has published many books for young readers. Science and environmental issues are among her favorite subjects. She lives in Oregon and enjoys observing the natural world while hiking, camping, and scuba diving.

brown jellyfish

9